I0877242

U.S. AIR FORCE

JEN BESEL

BLACK RABBIT BOOKS

Bolt Jr. is published by Black Rabbit Books
P.O. Box 227, Mankato, Minnesota, 56002
www.blackrabbitbooks.com
Copyright © 2023 Black Rabbit Books

Michael Sellner, designer and photo researcher

All rights reserved. No part of this book may be reproduced in any form without written permission from the publisher.

Names: Besel, Jennifer M., author.
Title: U.S. Air Force / by Jen Besel.
Other titles: United States Air Force
Description: Mankato, Minnesota : Black Rabbit Books, [2023] | Series: Bolt Jr. mighty military | Includes bibliographical references and index. |
Audience: Ages 6-8 | Audience: Grades K-1 |
Summary: "Engage young readers with information about the U.S. Air Force. From the group's fleet of aircraft to the people who fly and care for it, readers will learn all about how the U.S. Air Force protects the United States"– Provided by publisher.
Identifiers: LCCN 2021022163 (print) | LCCN 2021022164 (ebook) | ISBN 9781623107604 (hardcover) | ISBN 9781644665879 (paperback) | ISBN 9781623107666 (ebook)
Subjects: LCSH: United States. Air Force–Juvenile literature.
Classification: LCC UG633 .B45 2023 (print) | LCC UG633 (ebook) | DDC 358.400973–dc23
LC record available at https://lccn.loc.gov/2021022163
LC ebook record available at https://lccn.loc.gov/2021022164

Image Credits
Alamy: 508 collection, 23; Stocktrek Images, Inc., 1; Z2A1, 12; Dreamstime: Zoom-zoom, 10; Getty: Stocktrek Images, 18; MBDA Systems: Unknown, 21; Shutterstock: Dzmitry Kim, 3, 24; FabianGame, 15; Ivan Cholakov, 5; ledokolua, 14; Soonthorn Wongsaita, 4; Yaroslav Shkuro, 15; Sketchfab: skyhaven4259, 8–9; U.S. Air Force: Lt Col Leslie Pratt, 13; Sr. Airman Julianne Showalter, 6–7; Staff Sgt. Kenny Kennemer, 17; U.S. Air National Guard: Airman 1st Class Andrea A. S. Williamson, 11; U.S. Department of Defense: U.S. Air Force: 20; Wikipedia Commons: U.S. Air Force: Cover

Contents

CHAPTER 1

What Is the U.S. Air Force?

A plane flies above enemy land. An Air Force **pilot** drops bombs. The bombs will destroy weapons below.

pilot: a person who flies an airplane

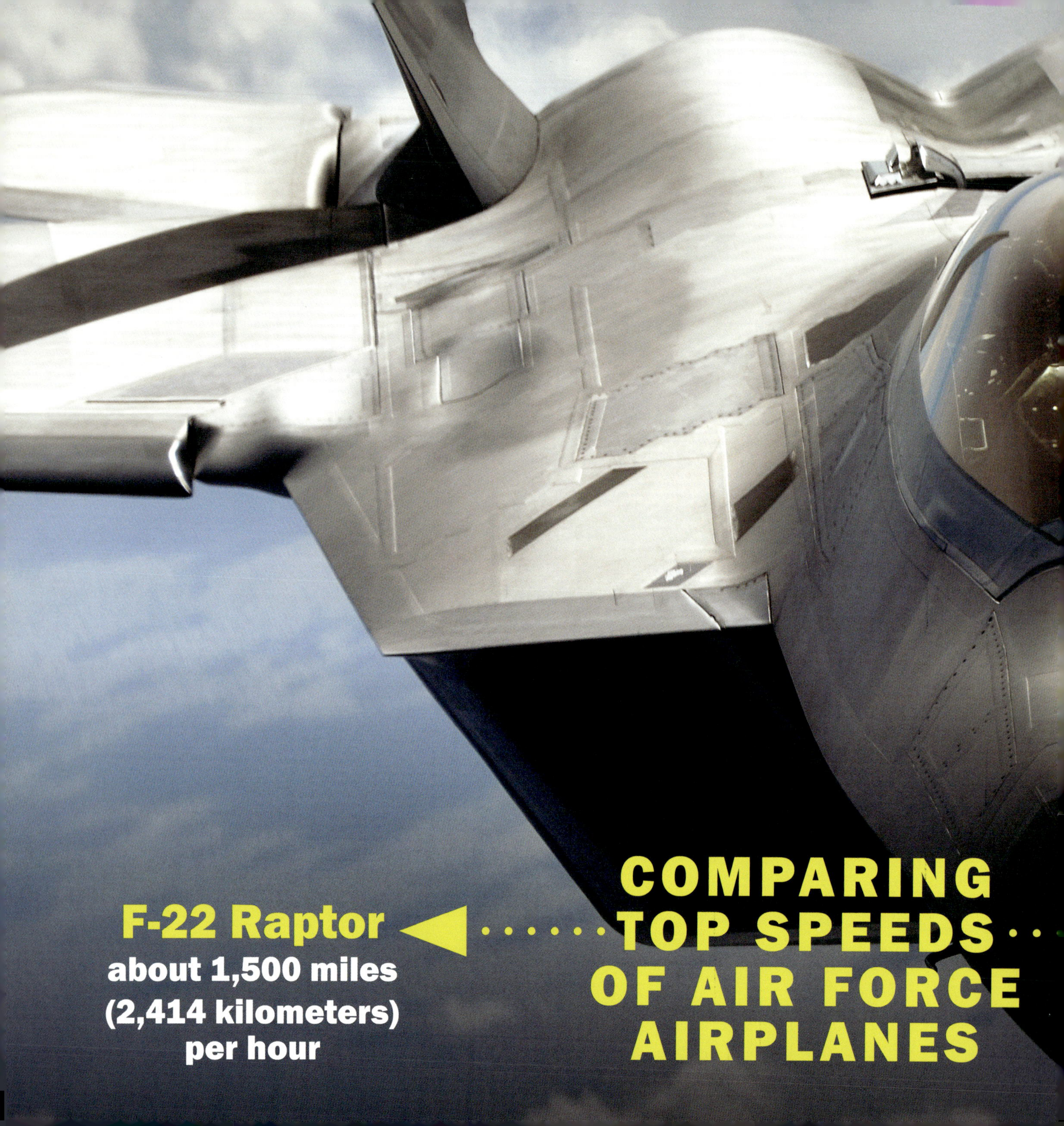

COMPARING TOP SPEEDS OF AIR FORCE AIRPLANES

F-22 Raptor
about 1,500 miles (2,414 kilometers) per hour

In the Sky

The Air Force is a part of the U.S. military. This group flies aircraft. All members of the group are called airmen. Airmen help people. They also build runways.

F-35 Lightning
about 1,200 miles (1,931 km) per hour

PARTS OF AN

F-22 Raptor

engines
wings

CHAPTER 2

Training

Airmen have a lot to learn. They start with **basic training**. Airmen build their strength. They learn to work together.

basic training: the first few weeks of learning for someone who has joined the military

FACT

The training lasts about eight weeks.

On the Job

After basic training, airmen get jobs. Some fly planes or **drones**. Others fix them. Some even study the weather.

drone: an unmanned aircraft or ship guided by remote control or onboard computer

Where Basic Training Happens

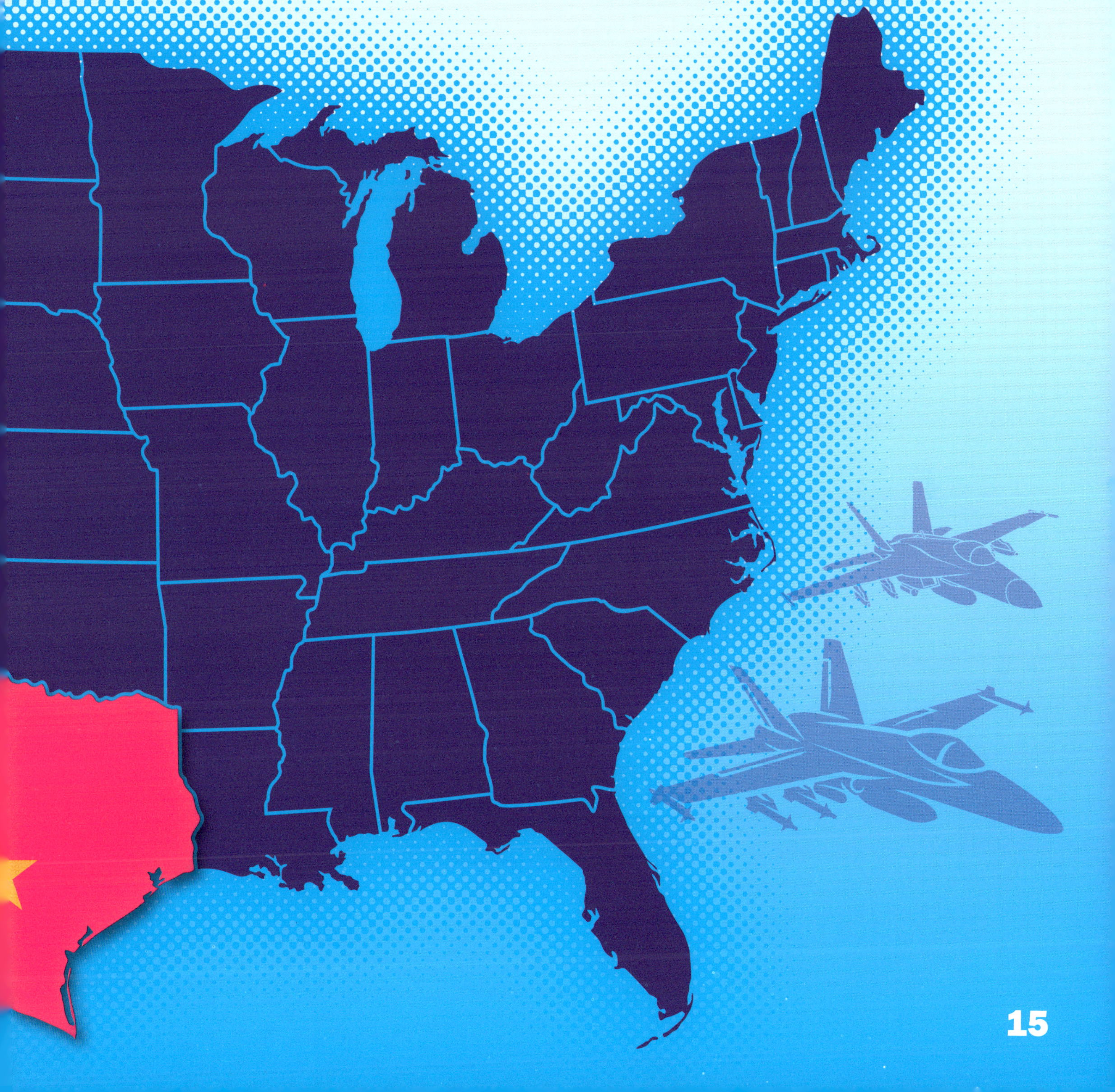

CHAPTER 3

Gear and Weapons

The Air Force has more than 5,000 aircraft. Some planes carry bombs. Choppers carry supplies. Drones spy on people.

Helmets

Pilots wear helmets. The helmets have microphones. Pilots talk to each other through them. Pilots also talk to people on the ground. They tell the airmen where to fly.

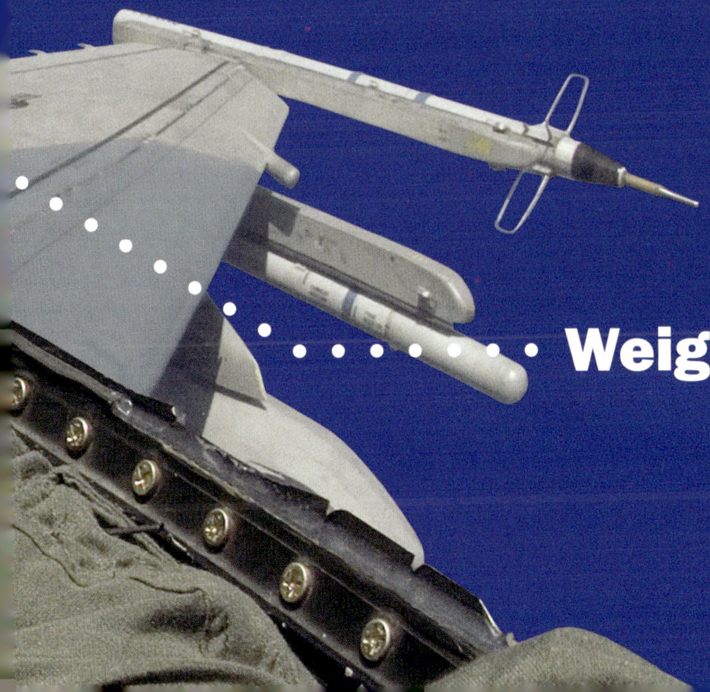

Weight of Pilot's Helmet

about 5 pounds (2.3 kilograms)

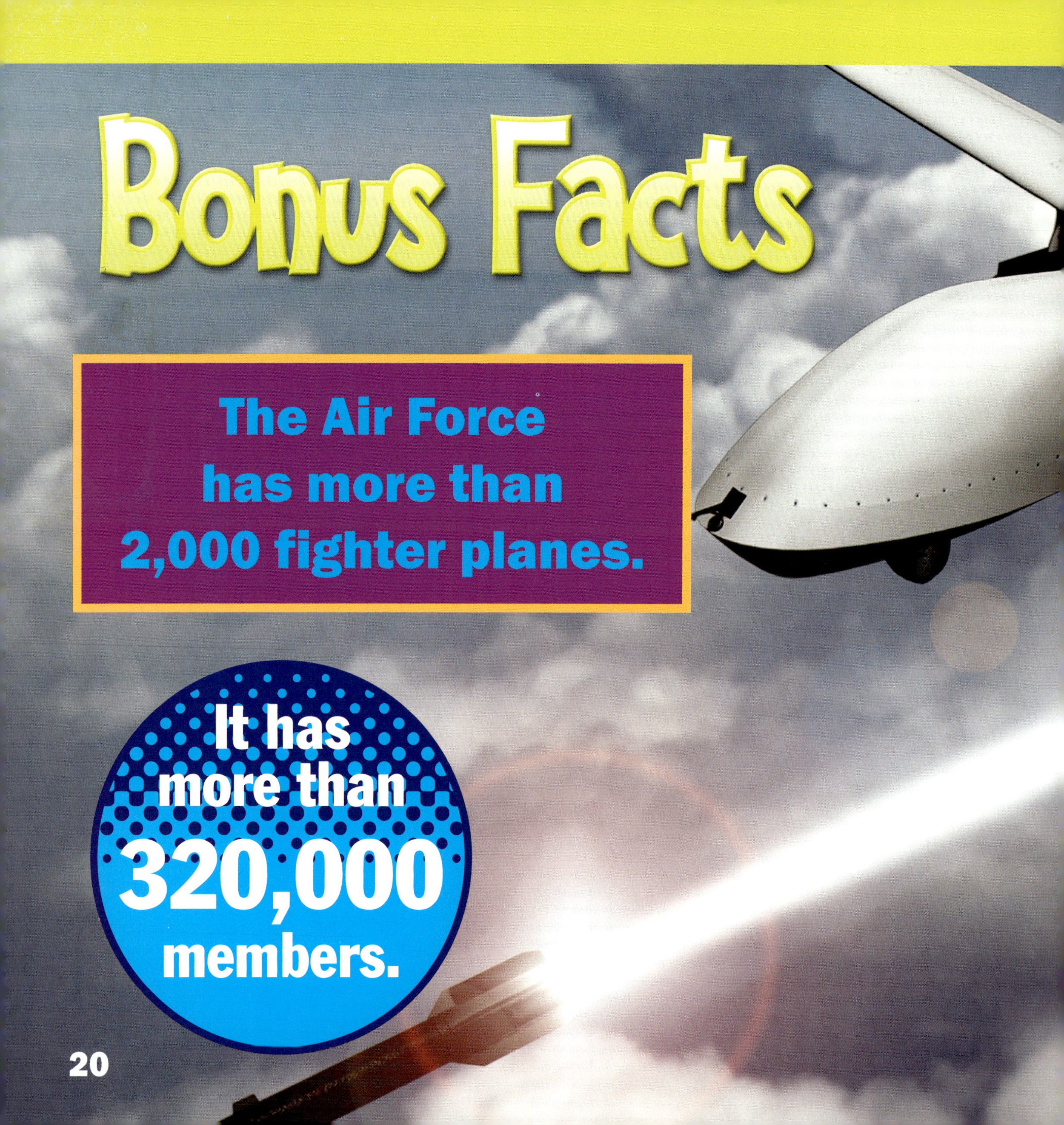

Bonus Facts

The Air Force has more than 2,000 fighter planes.

It has more than **320,000** members.

Airmen fly drones from the ground.

Some airmen stop **hackers.**

hacker: a person who sneaks into a computer system

READ MORE/WEBSITES

Billings, Tanner. *The U.S. Air Force.* Inside the U.S. Military. New York: Rosen Publishing, 2022.

Leed, Percy. *The U.S. Air Force in Action.* U.S. Military Branches. Minneapolis: Lerner Publications, 2023.

Ringstad, Arnold. *U.S. Air Force: Equipment and Vehicles.* U.S. Military and Equipment. Minneapolis: Kids Core, an Imprint of Abdo Publishing, 2022.

United States Air Force—YouTube Channel
www.youtube.com/user/AFBlueTube

U.S. Air Force
www.airforce.com/

GLOSSARY

basic training (BAY-sik TRAYN-ing)—the first few weeks of learning for someone who has joined the military

drone (DROHN)—an unmanned aircraft or ship guided by remote control or onboard computer

hacker (HAK-uhr)—a person who sneaks into a computer system

pilot (PY-lut)—a person who flies an airplane

INDEX